Math Twisters
Grade 7

By
CINDY BARDEN

COPYRIGHT © 2002 Mark Twain Media, Inc.

ISBN 1-58037-196-5

Printing No. CD-1542

Mark Twain Media, Inc., Publishers
Distributed by Carson-Dellosa Publishing Company, Inc.

Table of Contents

Introduction

Mark Twain Media, Inc., is pleased to present *Math Twisters*, a series of books designed to supplement your math curriculum by providing challenging and fun activities related to vital mathematical skills appropriate for each grade level.

Activities in this book reinforce skills appropriate for a seventh-grade math curriculum at home or in the classroom using National Council of Teachers of Mathematics (NCTM) guidelines to promote essential mathematical skills.

- **Problem-Solving** – Students learn to understand and investigate math through open-ended activities and extended problem-solving projects requiring a variety of strategies.

- **Communications** – Verbal and written skills are learned through math games and stories.

- **Reasoning** – Opportunities for inductive and deductive reasoning are provided through logic-based activities.

- **Connections** – Students learn to associate math skills with other subjects as well as the world outside the classroom through activities related to everyday math situations, history, and social studies.

- **Number Sense, Operations, and Computations** – Understanding of basic operations, number sense, and place value is achieved through activities that reinforce these concepts.

- **Patterns, Functions, and Relationships** – Students learn to collect, classify, organize, and interpret data and to explore mathematical patterns through activities that encourage the development and use of graphs, tables, and charts.

- **Algebra** – Activities are used as a method of understanding variables, expressions, and equations.

- **Statistics and Probability** – Exercises emphasize techniques to collect, describe, organize, analyze, evaluate, and interpret data.

- **Geometry** – Students will develop an understanding of geometry through activities that explore geometric figures, properties, spatial sense, and relationships.

Processes and formulas

For Quick Reference

Keep this guide in your math folder for quick reference in case you forget a process or formula.

Working With Fractions:

Add: Convert fractions to equivalent fractions with common denominators. Add numerators.

Subtract: Convert fractions to equivalent fractions with common denominators. Subtract numerators.

Multiply: Multiply the numerators. Multiply the denominators.

Divide: Invert the second fraction and multiply.

Working With Positive and Negative Numbers:

The product of two positive or two negative numbers is a positive number.

The product of a positive and a negative number is a negative number.

The quotient of two positive or two negative numbers is a positive number.

The quotient of a positive and a negative number is a negative number.

Formulas:

Perimeter of any geometric figure with straight sides: sum of all sides

Area of square or rectangle: *lw* (*l* = length and *w* = width)

Area of triangle: $\frac{1}{2}bh$ (*b* = base and *h* = height)

Area of a parallelogram: *bh* (*b* = base and *h* = height)

Area of circle: πr^2 (π = 3.14 and *r* = radius)

Circumference of a circle: πd (π = 3.14 and *d* = diameter)

Volume of a cube or rectangular prism: *lwh* (*l* = length, *w* = width, *h* = height)

Volume of a sphere: $\frac{4}{3}\pi r^3$ (π = 3.14 and *r* = radius)

Volume of a cylinder: *bh* (*b* = area of base and *h* = height)

Volume of a cone: $\frac{1}{3}\pi r^2 h$ (π = 3.14, *r* = radius, *h* = height)

Name: _____ Date: _____

Apply math to everyday activities

What Good Is Math?

Directions: Sometimes people wonder why they need to learn math. Math has many everyday applications, from determining the best price of products to balancing a checkbook. People use math to figure their income taxes, to determine how much lumber to buy, or to determine how much carpeting will cost. Give one specific example of how each math skill can be used in everyday life.

A.   Addition _____

B.   Subtraction _____

C.   Multiplication _____

D.   Division _____

E.   Adding, subtracting, multiplying, or dividing fractions

 _____

F.   Adding, subtracting, multiplying, or dividing decimals

 _____

G.   Changing fractions to decimals or decimals to fractions

 _____

H.   Finding the perimeter _____

I.   Finding the area _____

J.   Finding the volume _____

K.   Calculating percents _____

L.   Estimating _____

M.   Rounding _____

N.   Graphing _____

Name: _____　Date: _____

Review various operations

Historical Math

Directions: Read the following information and answer the questions.

In September 1833, Benjamin Day published the first successful penny newspaper, *The New York Sun*. Soon the paper's circulation rose to 30,000.

A.　At one cent each, how much did Benjamin Day receive each day for selling 30,000 copies?

_____

Even though 77,988 automobiles were registered in the United States by 1905, most people still considered horses more dependable. In 1909, a Model T Ford sold for $950. Henry Ford set up his first automobile assembly line for production of the Model T Ford in 1913. He paid his workers a fabulously high wage—$5 a day—and was still able to reduce the cost of making his cars.

B.　How much would a worker have made per year working six days a week for Henry Ford?

_____

By 1926, the price of a new Model T had been reduced to $290.

C.　What is the percent of reduction in the cost of the Model T between 1909 and 1913? (Round answer to nearest tenth of a percent.)

_____

A minimum wage of 40 cents per hour was established on July 12, 1933.

D.　How much would a worker have made in a year at minimum wage for 40 hours of work per week? _____

On July 22, 1933, aviator Wiley Post ended his first around-the-world flight. He averaged 83.67 mph during the trip, which lasted 7 days, 18 hours, and 15 minutes.

E.　How far did he travel? _____

The 1990 census was taken by a staff of over 315,000 and cost 2.6 billion dollars. The population of the United States in 1990 was 248,700,000.

F.　To the nearest cent, how much per person did it cost for the 1990 census?

_____

　　　　　　4

Name: _____ Date: _____

Find averages

Tackle This

Directions: The 11 offensive starters are pretty big guys, but the defensive players are even bigger. Answer the following questions about the team.

Offense	Height	Weight		Defense	Height	Weight
Bret	6′ 2″	225	Santana	6′ 5″	287
LeRoy	6′ 0″	204	Earl	6′ 4″	317
Antuan	6′ 1″	210	Bubba	6′ 6″	260
Bill	6′ 3″	205	Vonnie	6′ 5″	290
Ahman	6′ 0″	217	Gilbert	6′ 2″	339
Corey	6′ 1″	196	James	6′ 3″	266
Antonio	6′ 1″	198	Mike	6′ 5″	297
Frank	6′ 3″	305	Chad	6′ 5″	327
Earl	6′ 4″	317	Torrance	6′ 2″	255
Casey	6′ 1″	197	Bernardo	6′ 2″	250
Marco	6′ 4″	310	Rod	6′ 3″	320

		Offense	**Defense**
A.	Total weight	_____	_____
B.	Average weight (round to nearest pound)	_____	_____
C.	Average height (round to nearest inch)	_____	_____

D.   On the average, which group weighs more, the offense or defense? _____

E.   How much more? _____

F.   On the average, which group is taller, the offense or defense? _____

G.   How much taller? _____

5

Name: _____ Date: _____

Continue the patterns

Pascal's Triangle

Directions: Follow the pattern to complete the next six rows in Pascal's* Triangle.

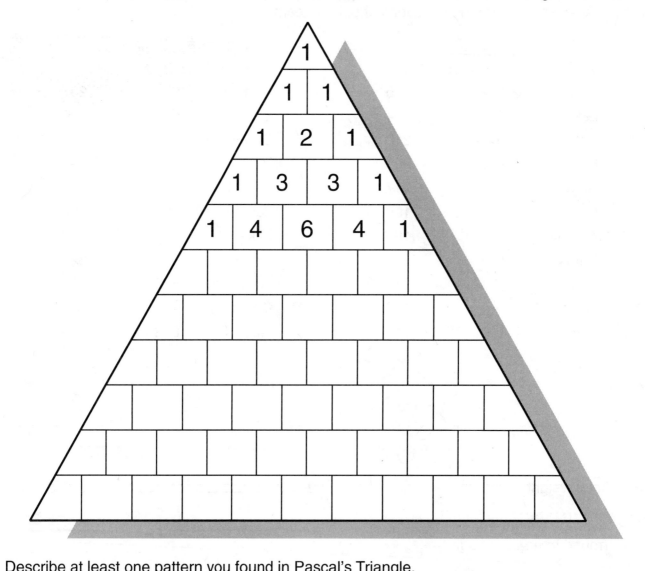

Describe at least one pattern you found in Pascal's Triangle.

_____

_____

_____

_____

* Blaise Pascal was a French scientist, philosopher, and mathematical prodigy. His contributions to mathematics include: the formulation of probability theory, the development of differential calculus, as well as Pascal's Law and Pascal's Triangle.

Name: _____ Date: _____

Compare and order integers

Brrrr!

On a January day, these temperatures were reported.

A.   Number the cities from 1 (coldest) to 5 (warmest).

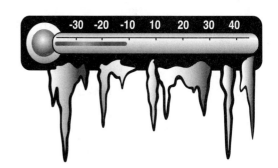

-11°F   Boston   _____
- 6°F   Chicago   _____
-14°F   Detroit   _____
-21°F   Minneapolis   _____
  3°F   Pittsburgh   _____

Compare. Write < (less than) or > (greater than) in the blanks.

B.   -4° ____ -2°   C.   -11° ____ -12°   D.   -6° ____ -12°

E.   -3° ____ 0°   F.   7° ____ -1°   G.   -5° ____ -7°

Write the integers in order from the coldest to the warmest.

H.   4   -7   2   -6   0   3

_____

I.   1   4   -9   6   -8   -12   0

_____

J.   5   -6   7   -8   4   -3   2

_____

Write the next three integers in the pattern.

K.   5   3   1   -1   ____ ____ ____   L.   -7   -5   -3   -1   ____ ____ ____

M.   10   5   0   -5   ____ ____ ____   N.   -20   -15   -10   -5   ____ ____ ____

Describe the weather on the coldest day you've ever experienced. How cold was it?

_____

_____

_____

_____

Name: _____ Date: _____

Calculate wages/overtime/vacation pay

Help Wanted

At the XYZ Factory, workers are paid by the hour. They receive time and a half for all hours over 40 in a week.

Directions: Calculate the weekly pay for these workers:

Name	Hourly Rate	Hours Worked	Gross Pay
Abby	$10.75	42	_____
Ben	$12.50	48	_____
Carlos	$15.15	45	_____
Dana	$11.90	40.5	_____
Eduardo	$13.48	44.5	_____

A.   Greg earns $14.50 per hour. He worked exactly the same number of hours each week for the past four weeks. His total pay, including overtime, for that period was $2,668.

How many hours per week did he work? _____

Vacation pay is based on gross pay for the previous year. One week of vacation is paid at 2% of last year's gross pay.

B.   Dana earned $25,644 last year. How much will she receive for one week of vacation?

_____

C.   Carlos earned $33,784 last year. He receives three weeks of vacation. What will his total vacation pay be this year?

_____

Workers at the XYZ Factory were given two options.
1) They could work four twelve-hour days one week and three twelve-hour days the next week

or
2) They could work five nine-hour days per week.

D.   Which option would you select? _____

E.   Why? _____

_____

_____

Name: _____ Date: _____

Track hypothetical investments

You have $10,000 to invest in the stock market for one week. Can you make money?

You can invest the money in any combination of stocks, with a minimum of two. The Internet is a great source of information about current stock prices. You can also find current stock prices in many metropolitan newspapers.

A.   Write the names of the stocks you selected on the first line. If you selected more than five, make a copy of this page to track all of your investments.

B.   Write the cost of each stock on the lines for Day 1.

C.   Write the closing price for each stock each day on the lines for Days 2 through 7.

Stock   _____ _____ _____ _____ _____

Day 1   _____ _____ _____ _____ _____

Day 2   _____ _____ _____ _____ _____

Day 3   _____ _____ _____ _____ _____

Day 4   _____ _____ _____ _____ _____

Day 5   _____ _____ _____ _____ _____

Day 6   _____ _____ _____ _____ _____

Day 7   _____ _____ _____ _____ _____

D.   When you cash in your portfolio at the end of one week, how much is each stock worth?

_____ _____ _____ _____ _____

E.   Which stock had the greatest return? _____

F.   Which stock had the lowest return? _____

G.   What is the total worth of your $10,000 investment after one week? _____

Name: _____ Date: _____

Solve equations using various operations

Body Math

Directions: Use the information below to answer the questions.

The average person's heart beats 103,680 times a day.

A.   How many heartbeats is that per minute? _____

Take your pulse for 15 seconds after exercising strenuously for 10 minutes. Multiply by four to get your rate per minute.

B.   What was your pulse rate? _____

Take your pulse for 15 seconds after resting for 10 minutes.

C.   What was your pulse rate? _____

Human blood travels about 61,320 miles in a year.

D.   How far does it travel in one day? _____

E.   In one hour? _____

Human hair grows at a rate of about 0.5 inches per month.

F.   If your hair were 5 inches long now, about how many months would it take for it to be 21 inches long?

_____

In May 1992, the hair of a woman in Massachusetts measured 12 feet 2 inches long.

G.   How long would it take to grow hair that long? _____

Fingernails grow at a rate of about 1/25 inch per week—four times faster than toenails. In 1992, a man in India had one fingernail that was 40 inches long.

H.   About how long did it take for that fingernail to grow? _____

Did You Know? It takes only 17 facial muscles to smile, but it takes 43 muscles to frown.

Name: _____ Date: _____

Compare prices

Then and Now

Everything was cheaper in the good old days, right? Not necessarily.

Microwave ovens were first introduced in Mansfield, Ohio, by the Tappan Company on October 25, 1955. The price tag: $1,200.

A.   Check the prices of microwaves at stores, in newspaper ads, or online. Fill in the information for three different models. Compare the prices then and now.

Brand name	Size	Price	Difference between now and 1955
_____	_____	_____	_____
_____	_____	_____	_____
_____	_____	_____	_____

In 1945, Gimbels Department Store in New York City was the first to sell commercially-made ballpoint pens. The pens sold for $12.50 each.

B.   Check the price of a disposable ballpoint pen today. _____

How much less does a ballpoint pen cost today than it did in 1945? _____

It cost $10 to see the first talking picture, "Don Juan," starring John Barrymore. The black and white movie was first shown at New York's Warner Theater on August 6, 1926.

C.   How much more (or less) does a ticket to a movie cost today where you live than in 1926? _____

Imagine paying $850 for a new car! That's what it cost to buy the new Model T introduced by Henry Ford on October 1, 1908.

D.   Check the price on three new Ford vehicles, like a car, a van, and a pick-up. Look for the lowest prices you can find (stripped-down models, no extras.) Compare the prices then and now.

Type of vehicle	Price	Difference between 1908 and now
_____	_____	_____
_____	_____	_____
_____	_____	_____

Name: _____   Date: _____

Complete Venn diagrams

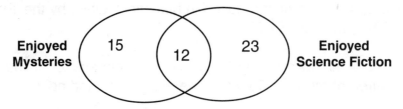
Surveys Show

A survey showed that out of 50 people, 27 said they enjoyed mysteries, and 35 said they liked science fiction books.

Enjoyed Mysteries   15   12   23   **Enjoyed Science Fiction**

Use the Venn diagram to answer the questions.

A.   How many people enjoyed both mysteries and science fiction? _____

B.   How many enjoyed only mysteries? _____

C.   How many enjoyed only science fiction? _____

Of 84 people surveyed, 47 said they liked to eat popcorn while watching a movie, and 56 said they liked to eat chips and dip while watching a movie.

D.   Complete the Venn diagram to show the results of this survey.

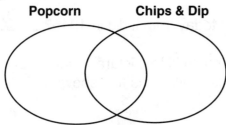
Popcorn   **Chips & Dip**

E.   How many of the people surveyed enjoyed both popcorn and chips and dip while watching a movie?

_____

Of 100 people surveyed, 53% said they owned a pair of black shoes, and 67% said they owned a pair of running shoes.

F.   Complete the Venn diagram to show the results of this survey. Include labels on the diagram.

Multiply and divide fractions and decimals

Directions: You will need pencils and scrap paper to play this version of Tic-Tac-Go. It is played like regular tic-tac-toe. However, before you can put an "X" or an "O" on the board, you must solve the equation. Players check each other's answers. If incorrect, the player skips a turn.

A.

$3.5 \times 6.2 =$	$8.1 \times 7.4 =$	$0.71 \times 7.01 =$
$9\frac{1}{3} \times 7.3 =$	$5.7 \times 2\frac{1}{3} =$	$6.2 \times 4\frac{1}{2} =$
$6\frac{2}{3} \times 3.9 =$	$4.02 \times 5\frac{1}{8} =$	$8.47 \times 6.2 =$

B.

$10^4 =$	$4^3 =$	$5^2 =$
$2^6 =$	$3^6 =$	$10^3 =$
$4^4 =$	$5^3 =$	$6^3 =$

C.

$3\frac{1}{2} \times 6 =$	$2\frac{1}{4} \times 3\frac{1}{2} =$	$4 \times \frac{4}{9} =$
$8 \times 2\frac{1}{2} =$	$4 \times \frac{4}{16} =$	$5 \times \frac{7}{3} =$
$7 \times \frac{4}{9} =$	$3\frac{2}{3} \times 4\frac{3}{4} =$	$9\frac{2}{5} \times 1\frac{1}{2} =$

D.

$8 \div 2\frac{1}{2} =$	$9 \div \frac{1}{3} =$	$7 \div \frac{2}{5} =$
$6 \div \frac{3}{7} =$	$3 \div \frac{5}{6} =$	$5 \div \frac{2}{9} =$
$2 \div \frac{3}{8} =$	$4 \div \frac{1}{16} =$	$12 \div \frac{3}{5} =$

Reduce to lowest terms

E.

$\frac{17}{4} =$	$\frac{92}{20} =$	$\frac{13}{9} =$
$\frac{19}{3} =$	$\frac{24}{7} =$	$\frac{26}{8} =$
$\frac{42}{5} =$	$\frac{39}{6} =$	$\frac{41}{2} =$

F.

$3\frac{16}{4} =$	$2\frac{9}{5} =$	$4\frac{18}{3} =$
$7\frac{14}{2} =$	$5\frac{9}{5} =$	$9\frac{7}{2} =$
$1\frac{24}{3} =$	$6\frac{8}{3} =$	$8\frac{48}{12} =$

Name: _____ Date: _____

Collect and organize data

Collect Data

A. Write the total number of pages in each of five textbooks you use.

	Subject	**Number of pages**
Book 1	_____	_____
Book 2	_____	_____
Book 3	_____	_____
Book 4	_____	_____
Book 5	_____	_____

What is the average number of pages?

_____

B. Measure five ice cubes in centimeters. Find the volume of each. Record the data below.

	Dimensions	**Volume**	**Amount of water**
Cube 1	_____	_____	_____
Cube 2	_____	_____	_____
Cube 3	_____	_____	_____
Cube 4	_____	_____	_____
Cube 5	_____	_____	_____

C. Melt the ice cubes in separate containers. Measure the amount of water in each container, and record it on the lines above.

D. Hold an apple-, orange-, or potato-peeling contest. Measure the length of the 10 longest unbroken peelings. Create a table and a graph at the bottom of this page or on your own paper showing the results. You may use a bar graph, line graph, pictograph, circle graph, or any other type to illustrate the data.

Name: _____ Date: _____

Create a graph/Calculate percents

For this project you will need several bags of M & Ms™.

A.　Work with a group. Sort the candies by colors. Fill in the table.

Color	Number

B.　How many candies were there in all? _____

C.　Make a graph to show how many candies of each color you have.

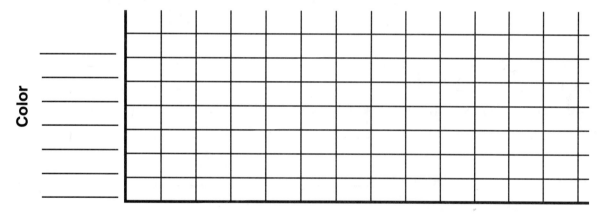

Write the answers. Round to the nearest whole percent.

D.　What percent of the candies were yellow? _____

E.　What percent of the candies were green? _____

F.　What percent of the candies were red? _____

G.　What percent of the candies were brown? _____

H.　What percent of the candies were orange? _____

I.　What percent of the candies were blue? _____

J.　What percent of the candies were purple? _____

When you finish, enjoy the candy.

Name: _____ Date: _____

Calculate equivalent fractions, decimals, and percents

Music to My Ears

The high school band has 100 members. They include:

Woodwind Section
22 B-flat clarinet players
4   bass clarinet players
11 flute players
4   piccolo players
2   oboe players
1   bassoon player
7   alto saxophone players
5   tenor saxophone players
2   bass saxophone players

Brass Section
6   French horn players
9   trumpet players
4   cornet players
8   trombone players
3   tuba players

Percussion Section
5   snare drummers
2   bass drummers
1   cymbal player
3   timpani players
1   marimba/xylophone player

Directions: Show the proportion of the band members who play the instruments listed below. Write each answer as a decimal, a percent, and a fraction. Reduce all fractions to lowest terms.

A.   Members who play tenor saxophone

   Percent: _____ Decimal: _____ Fraction: _____

B.   Members who play French horn

   Percent: _____ Decimal: _____ Fraction: _____

C.   Cymbal players

   Percent: _____ Decimal: _____ Fraction: _____

D.   Trumpet and cornet players

   Percent: _____ Decimal: _____ Fraction: _____

E.   Trombone and tuba players

   Percent: _____ Decimal: _____ Fraction: _____

F.   Tenor and bass saxophone players

   Percent: _____ Decimal: _____ Fraction: _____

G.   Members who play woodwind instruments

   Percent: _____ Decimal: _____ Fraction: _____

H.   Members who play brass instruments

   Percent: _____ Decimal: _____ Fraction: _____

I.   Members of the percussion section

   Percent: _____ Decimal: _____ Fraction: _____

Name: _____ Date: _____

Use logic and deductive reasoning

Party Time

Directions: Four friends worked together to throw an end-of-the-summer party. From the clues given, determine each person's last name, the job each performed, and the color of T-shirt each person wore. Write "N" for no or "Y" for yes in the grid. Use a pencil in case you need to erase.

1. The one who made the food wore a red T-shirt and had a first name two letters longer than the one who sent the invitations.

2. The person whose last name was Carlson wore a green T-shirt.

3. Carlos's last name was Juarez, but Rob's wasn't O'Brien.

4. The person whose last name was Edwards decorated for the party and had a first name that was one letter longer than the person who wore the blue T-shirt.

	Juarez	Carlson	Edwards	O'Brien	Red	Blue	Green	Purple	Made food	Invitations	Decorated	Cleaned up
Rob												
Tina												
Maria												
Carlos												
Made food					Y	N	N	N				
Invitations					N							
Decorated					N							
Cleaned up					N							
Red												
Blue												
Green												
Purple												

First Name	Last Name	Job	Color T-shirt
Rob			
Tina			
Maria			
Carlos			

Locate points on a grid

What's the Point?

You will need pencils and two dice for this two-player game.

Directions:

1. Players take turns rolling the two dice, one at a time. The number on the first die is the number for the x-axis. The number on the second die is the number for the y-axis.

2. Each player marks the point on his or her grid indicated by the two numbers on the dice. Players check each other. If the point is incorrect, the player erases and skips that turn.

3. Play continues until one player correctly marks all points on his or her grid.

Player 1

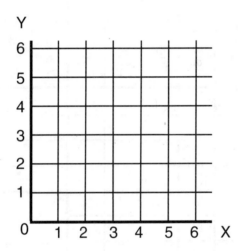

Player 2

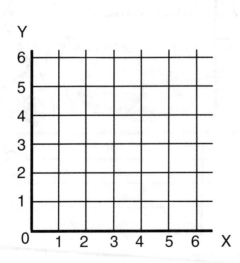

Name: _____ Date: _____

Explore combinations

Decisions, Decisions

Josh's parents plan to buy two items from his "wish list" for his birthday, but they haven't decided which two. Before they make up their minds, they want to know all their options.

Josh's Wish List

A Computer game
B Basketball
C Hockey stick
D Skateboard
E Ticket to an NFL game

F Music CD
G Baseball glove
H Scanner
I Microscope

Directions: Complete the chart to show all the options for two gifts from Josh's wish list. Use the letters of the items.

A + B	B + C	C + ___	D + ___	E + ___	F + ___	G + ___	___ + ___	___ + ___
A + C	___ + ___	C + ___	D + ___	E + ___	F + ___	G + ___	___ + ___	___ + ___
A + D	___ + ___	C + ___	D + ___	E + ___	F + ___	G + ___	___ + ___	___ + ___
A + ___	___ + ___	C + ___	D + ___	E + ___	F + ___	G + ___	___ + ___	___ + ___
A + ___	___ + ___	C + ___	D + ___	E + ___	F + ___	G + ___	___ + ___	___ + ___
A + ___	___ + ___	C + ___	D + ___	E + ___	F + ___	G + ___	___ + ___	___ + ___
A + ___	___ + ___	C + ___	D + ___	E + ___	F + ___	G + ___	___ + ___	___ + ___
A + ___	___ + ___	C + ___	D + ___	E + ___	F + ___	G + ___	___ + ___	___ + ___
A + ___	___ + ___	C + ___	D + ___	E + ___	F + ___	G + ___	___ + ___	___ + ___

HINT: You do not need all the spaces above. Use only the ones you need.

A. How many different options do Josh's parents have? _____

B. What pattern did you notice as you filled in the chart? _____

_____

_____

C. Which two gifts would you most like? _____

_____

Name: _____ Date: _____

Explore permutations

Stop and Smell the Roses

Lisa has room to plant four rose bushes near the front door of her house. She wants to plant four different colored rose bushes: white, red, peach, and lavender.

Directions: Show all possible ways Lisa could arrange the roses. You can abbreviate by using "W" for white, "R" for red, "P" for peach, and "L" for lavender.

W R P L	R P L W		

Use as many of the spaces on the chart as you need. Add more spaces if needed.

A.   How many ways can Lisa arrange the four rose bushes? _____

Name: _____   Date: _____

Use deductive reasoning/Algebra/Mixed operations

Brain Teasers

Directions: Find the answers. **Hint:** It might help to draw pictures on scrap paper as you investigate possible solutions.

A.   What three consecutive numbers have a sum of 150? _____ _____ _____

B.   Scott and Tyrone play basketball. The sum of their heights is 13 feet, 4 inches. Scott is 2 inches taller than Tyrone. How tall is each player?

Scott: _____   Tyrone: _____

C.   There are three children in the Heimerdinger family: Harry, Henry, and Hannah. They are 2, 6, and 8 years old. Harry is older than Henry. Henry was born when Hannah was six years old.

How old is each child? Harry: _____ Henry: _____ Hannah: _____

D.   On a field trip, two school buses drove in front of a school bus and two school buses drove behind a school bus in a single-file line.

What is the fewest number of school buses that could be driving together on the trip?

_____

E.   Each school bus held a maximum of 48 passengers.

Bus 1 was $\frac{2}{3}$ full.

Bus 2 was $\frac{1}{4}$ empty.

Bus 3 had 87.5% of the maximum number of passengers.

How many passengers were on the three buses? _____

F.   Jason's grandfather is four times older than he is. The sum of Jason's age and his grandfather's age is 80.

How old is Jason? _____

G.   At a farmer's market, a bushel of potatoes cost $10, a 50-pound bag of onions cost $5, and pumpkins were $1 each.

How many different combinations of vegetables could you buy if you spent $26? _____

Name: _____ Date: _____

Review properties of lines

What's My Line?

Review

Line: A set of points that extends without end in opposite directions.

A B C

Line segment: Part of a line with two endpoints.

M   N

Ray: A part of a line with one endpoint.

G   H

Intersecting lines: Lines that cross each other.

Parallel lines: Lines on the same plane that never intersect.

Perpendicular lines: Two lines that form a right angle.

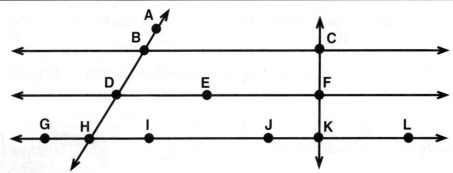

Directions: Use the diagram above to answer the following questions. Use the symbols for lines, line segments, and rays in your answers.

A. Give an example of a line segment from the diagram. _____

B. Give an example of a line from the diagram. _____

C. Give an example of a ray from the diagram. _____

For each pair, write intersecting, parallel, or perpendicular.

D. $\overleftrightarrow{BH}$ and $\overleftrightarrow{GI}$ _____

E. $\overleftrightarrow{CK}$ and $\overleftrightarrow{JL}$ _____

F. $\overrightarrow{GI}$ and $\overleftrightarrow{DF}$ _____

G. $\overrightarrow{BC}$ and $\overleftrightarrow{BH}$ _____

22

Name: _____ Date: _____

Calculate radius, diameter, and circumference

Around and Around

$$C = \pi \times d \text{ or } C = \pi \times 2r$$

C = circumference   **d** = diameter   **r** = radius $\pi = 3.14$

Directions: Round answers to the nearest tenth.

On September 3, 1970, a 1.67-pound hailstone fell in Coffeyville, Texas. The diameter of this huge hailstone was 7.5 inches!

A. What was its circumference? _____

Earth's moon has an average diameter of 2,159.3 miles.

B. What is its average circumference? _____

C. What is its average radius? _____

The first Ferris wheel, named after its constructor George Ferris, was erected in 1893 at the Chicago World's Fair. It had a diameter of 250 feet and carried 36 cars, each capable of holding 60 people.

D. What was its circumference? _____

E. How many people could ride at one time? _____

F. How far, in miles, would one car on the Ferris wheel travel in 10 revolutions? _____

A Ferris wheel erected in Yokohama, Japan, has a diameter of 328 feet.

G. What is its circumference? _____

H. What is the difference in circumference between the Ferris wheel in Japan and the one constructed in 1893?
_____

In June 1985, a woman in Fresno, California, set a world record for blowing a bubble-gum bubble with a radius of 11 inches.

I. What was the circumference of the bubble? _____

A clock in Japan has a face 101 feet in diameter. The minute hand is 41 feet long.

J. How much shorter is the minute hand than the radius of the clock? _____

K. What is the circumference of this clock? _____

Name: _____ Date: _____

Write algebraic expressions

Let *t* Represent Temperature

Directions: Write an algebraic expression for each statement. Use the variables given.

　　　Example:　　Let *c* represent the cost of a pizza. Write the algebraic expression that represents the cost of three pizzas.　　*c* * 3　or　3*c*

A.　Let *t* represent the temperature at 6 A.M.
　　Write the algebraic expression that represents the temperature after it rises 21 degrees.

　　_____

B.　Let *c* represent the cost of an ice cream cone.
　　Write the algebraic expression that represents the cost of six ice cream cones.

　　_____

C.　Let *n* represent any even number.
　　Write the algebraic expression that represents the next larger even number.

　　_____

D.　Let *w* represent any whole number.
　　Write the algebraic expressions that represent the next three whole numbers.

　　_____

E.　Let *s* represent the speed of a car in miles per hour.
　　Write the algebraic expression that represents a decrease in speed of 20 mph.

　　_____

F.　Let *h* represent the height of a hot air balloon.
　　Write the algebraic expression that represents a 25% increase in height.

　　_____

G.　Let *s* represent the number of seconds in one day.
　　Write the algebraic expression that represents the number of seconds in one week.

　　_____

H.　Let *m* represent the price of a meal.
　　Write the algebraic expression that represents a 15% tip.

　　_____

　　　　　　　24

Name: _____ Date: _____

Write true equations

To Tell the Truth

Change one number to make each equation true. Rewrite the equations.

A. $7 + x = x + 5$ _____

B. $98 + (-48) = 52$ _____

C. $15 * 3 = 60$ _____

D. $\frac{7}{3} = 2\frac{1}{2}$ _____

Write two more untrue equations like these. Trade papers with a classmate and solve.

E. _____ _____

F. _____ _____

Change one operational sign to make each equation true. Rewrite the equations.

G. $9 * 5 = 14$ _____

H. $(14 + 21) - 7 * 6 = 77$ _____

I. $\dfrac{(54 + 6) * 2}{8} = 12$ _____

J. $(7.3 * 3) * (6.4 * 2.1) = 8.46$ _____

Write two more untrue equations like these. Trade papers with a classmate and solve.

K. _____ _____

L. _____ _____

M. In your own words, define a true equation. _____

_____

Solve addition and subtraction equations with variables

X Marks the Spot

Each player needs a different colored pencil and one die.

Directions:

1. Both players roll one die. The player with the higher number selects any equation to solve. If correct, the player colors in that space. Players check each other's answers.

2. If both players shake the same number, both shake again.

3. Only the player with the higher number takes a turn each round.

4. The player with the most spaces colored is the winner.

$a + 14 = 21$	$d - 5 = 72$
$i - 0 = 935$	$7 + p = 71$
$c + 1 = 101$	$14 - y = 9$
$71 - m = 48$	$33 + j = 31$
$68 - o = 45$	$36 - s = 21$
$e + 15 = 45$	$t + 75 = 116$
$k + 11 = 365$	$4 + b = 19$
$p + 21 = 47$	$17 - r = 1$
$z - 3 = 76$	$n + 22 = 622$
$x + 7 = 11$	$h + 32 = 104$
$13 - w = 9$	$11 - f = 4$
$q + 50 = 155$	$l - 19 = 11$
$x - 10 = 47$	$100 - r = 4$
$5 - y = 3$	$u - 2 = 34$
$g - 19 = 91$	$8 + v = 58$
$3 + z = 41$	$16 - m = 3$

Name: _____ Date: _____

Construct triangles/Measure angles

Triangle Tango

A **right triangle** has one 90° angle.

A. Use a protractor. Draw a right triangle.

B. Label the three angles on your drawing: L, M, and N.

C. Measure the angles. L = _____°   M = _____°   N = _____°

D. What is the sum of the three angles? _____°

An **acute triangle** has all angles less than 90°.

E. Use a protractor. Draw an acute triangle.

F. Label the three angles on your drawing: X, Y, and Z.

G. Measure the angles. X = _____°   Y = _____°   Z = _____°

H. What is the sum of the three angles? _____°

An **obtuse triangle** has one angle greater than 90° and less than 180°.

I. Use a protractor. Draw an obtuse triangle.

J. Label the three angles on your drawing: G, H, and I.

K. Measure the angles. G = _____°   H = _____°   I = _____°

L. What is the sum of the three angles? _____°

The sum of the three angles of a triangle is always 180°.

Name: _____ Date: _____

Write and solve equations

World Records

Directions: Write the equation and circle the solution for each world record story. Round answers to the nearest tenth.

Pizza lovers in Havana, Florida, created a pizza covering 10,057 square feet in 1991, but it didn't set the record. An even larger one had been made the year before in Norwood, South Africa. That monster pizza measured 122.66 feet in diameter.

A.   What was the area of this monster pizza?

Equation: _____

B.   How much larger was it than the one made in Florida?

Equation: _____

A high temperature of 134° F was recorded in Death Valley, California, on July 10, 1913. A record low of -79.8° was set at Prospect Creek, Alaska, on June 23, 1971.

C.   How much difference was there between these two temperatures?

Equation: _____

A record-breaking apple pie contained over 600 bushels of apples and weighed 30,115 pounds. Baked in a 40- by 23-foot dish, this super pie was made in England.

D.   What was the area of the dish?

Equation: _____

E.   What was the perimeter of the dish?

Equation: _____

In 1976, Kathy Wafler peeled a 20-ounce apple in one long, unbroken peel that was 172 feet, 4 inches long.

F.   How long was the peeling in inches?

Equation: _____

G.   How long was it in yards?

Equation: _____

Did You Know? Taco Tico of Nebraska, Inc., set a record in 1991 for the world's largest burrito: 1,597 feet, 9 inches long. Made from 2,557 tortillas, this colossal burrito contained 607 pounds of refried beans and 75.75 pounds of shredded cheese!

Name: _____ Date: _____

Write and solve equations

Write Your Own Math Stories

Directions: Write any name between the square brackets [ ]. Fill in the parentheses ( ) with any number. Write the answers on the blanks.

A.   [   ] spent ($) at the music store. She gave the clerk a ($) bill. The clerk gave her $ _____ back.

B.   An alien traveled (   ) miles in his space ship at (   ) miles per second. It took the alien _____ to complete his journey.

C.   [   ] and (   ) friends went to a movie. Tickets cost ($) each. They spent _____ for tickets.

D.   [   ] filled three different sized containers with tomato juice. One container held (   ) ounces. The second held (   ) ounces, and the third held (   ) ounces.
     [   ] had _____ ounces of tomato juice in all.

E.   On Monday, the humane shelter had (   ) kittens and (   ) puppies. (   ) more kittens and (   ) more puppies were brought in that day. People adopted (   ) kittens and (   ) puppies. At the end of the day, there were _____ kittens and _____ puppies at the humane shelter.

F.   [   ] worked for (   ) hours shoveling snow and earned ($) per hour for a total of $ _____.

G.   [   ] had (   ) football cards. She gave (   ) cards to each of (   ) friends. She gave away a total of _____ cards.

H.   A large pizza was cut into (   ) pieces. [   ] ate (   ) pieces. _____% of the pizza was left.

I.   [   ] left for the library at (   ) A.M. and arrived (   ) hours and (   ) minutes later. The time was then _____.

J.   [   ]'s room was (   ) feet (   ) inches long and (   ) feet (   ) inches wide. The area of the room was _____.

K.   The tire on [   ]'s bike had a radius of (   ) inches. The circumference of the tire was _____.

29

Solve multiplication equations with one variable

Alien Invasion

Each player needs a die and a different colored pencil.

Directions:

1. Each player rolls one die. The player with the higher number selects any UFO and solves the equation. If correct, that player colors in that UFO. Players check each other's answers.

2. If both players roll the same number, both roll again.

3. Only the player with the higher number takes a turn each round.

4. The player with the most UFOs wins.

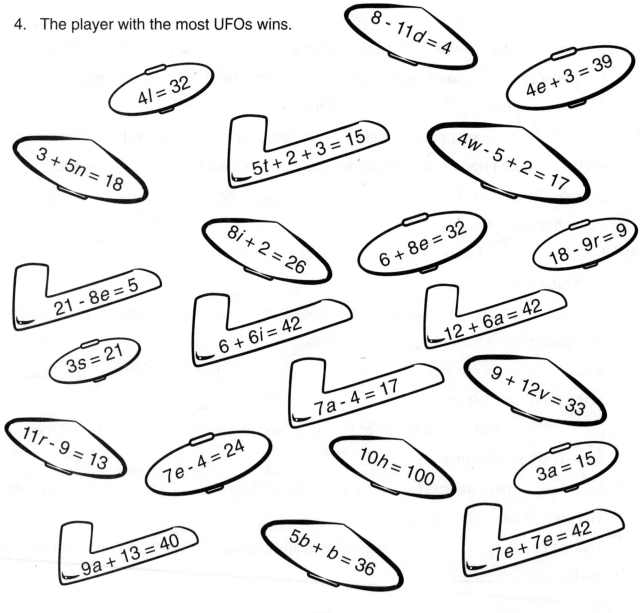

Name: _____ Date: _____

Compare types of triangles

More Triangles

An **equilateral triangle** has all sides and angles congruent.

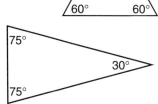

An **isosceles triangle** has two congruent angles and two congruent sides.

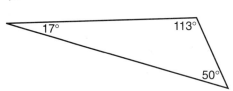

A **scalene triangle** has no congruent sides or congruent angles.

A. Can an equilateral triangle ever be a right triangle? _____

Why or why not? _____

_____

B. Can a right triangle ever be a scalene triangle? _____

Use a drawing to show your answer.

C. Use a protractor. Draw an example of an obtuse triangle that is also a scalene triangle. Label the three angles and the number of degrees of each angle.

D. Use a protractor. Draw an acute triangle that is also an equilateral triangle. Label the three angles and the number of degrees of each angle.

Name: _____ Date: _____

Creative thinking with geometric shapes

Swim Ballet

Members of the Swan's Ballet Swimming Club are planning their next program. As part of the show, they want to form a solid equilateral triangle of swimmers, then rearrange themselves into a solid square in the pool using the same number of swimmers.

An example of a solid equilateral triangle:

An example of a solid square:

A.   What is the minimum number of swimmers needed to form a solid equilateral triangle that could rearrange themselves into a solid square?

_____

B.   Can you form a solid square with 81 checkers without having any left over? Why or why not?

_____

_____

C.   Draw a diagram to show how you could use all 81 checkers to form several solid squares with no checkers left over.

D.   Can you form a solid equilateral triangle with 81 checkers without having any left over? Why or why not?

_____

_____

Solve division equations with one variable

Reach for the Stars

You will need four dice, colored pencils, and paper. Each player needs a different colored pencil.

Directions:

1. Players decide how many dice to roll on each turn: one, two, three, or four. The total number on the dice represents the number of the star. Each player must solve the equation on the star with the corresponding number. If correct, the player colors in that star.

2. Players check each other's answers. If a star has already been colored, that player skips a turn. The player with the most stars is the winner.

1. $\dfrac{a}{2} = 8$

2. $\dfrac{b}{7} = 1$

3. $\dfrac{12}{c} = 3$

4. $\dfrac{11}{d} = 2$

5. $\dfrac{e}{9} = \dfrac{1}{3}$

6. $\dfrac{8}{f} = 2$

7. $\dfrac{6}{g} = \dfrac{3}{5}$

8. $\dfrac{7}{h} = 3$

9. $\dfrac{i}{5} = 5$

10. $\dfrac{q}{8} = 7$

11. $\dfrac{10}{n} = 100$

12. $\dfrac{l}{5} = 2$

13. $\dfrac{3}{k} = 1$

14. $\dfrac{j}{4} = 4$

15. $\dfrac{2}{m} = 12$

16. $\dfrac{4}{s} = \dfrac{1}{4}$

17. $\dfrac{p}{10} = 3$

18. $\dfrac{7}{r} = \dfrac{1}{7}$

19. $\dfrac{u}{5} = 25$

20. $\dfrac{3}{x} = \dfrac{1}{3}$

21. $\dfrac{9}{w} = 3$

22. $\dfrac{y}{19} = 2$

23. $\dfrac{v}{6} = 6$

24. $\dfrac{11}{t} = 11$

Name: _____ Date: _____

Draw geometric figures

You have seen some geometric shapes hundreds, even thousands of times, but can you remember them exactly?

Directions: Draw the shapes. When you finish, compare your drawings to the actual sizes of the real items. Use your own paper if you need more room.

A.   Draw circles the size of a penny, nickel, dime, and quarter.

      penny   nickel   dime   quarter

B.   Draw a circle the size of the
      cap on a gallon of milk.

C.   Draw a circle the size of the
      bottom of a soda can.

D.   Draw a square the size of a
      key on a computer keyboard.

E.   Draw a line the length of your
      foot without a shoe.

F.   Draw a rectangle the size of the
      calculator you use most often.

G.   Draw a circle the size of the base of
      a light bulb.

Check yourself by comparing the actual items with your drawings.

Grade yourself on how well you did. My grade: _____

Name: _____ Date: _____

Solve for variables

Think It Through

A.   What are the next three numbers in the pattern?

       1   1   2   3   5   8   13   _____   _____   _____

B.   Jake has eight coins in his pocket worth a total of 74 cents. His sister wants to borrow exactly 30 cents.

    Can Jake lend her exactly 30 cents? _____ Why or why not? _____

    _____

C.   There are more than 30, but less than 50 monkeys at the zoo. When grouped in pairs, no monkey is alone. When in groups of five, there are no monkeys left over.

    How many monkeys are at the zoo? _____

D.   It costs $90 for one blue boomadinger and one red boomadinger. Three blue boomadingers and two red boomadingers cost $220.

    How much does one blue boomadinger cost? _____

E.   At the annual boomadinger sale, two small boomadingers and one large boomadinger cost the same as four small boomadingers.

    If a small boomadinger costs $40, what does a large boomadinger cost? _____

F.   Ahmad saved $1 bills and $5 bills in a shoe box. He had a total of 95 bills worth $275.

    How many ones and how many fives did he have? _____

Shape Up

Arrange toothpicks to match the shapes below.

A.   Move four toothpicks to change the six equilateral triangles into three equilateral triangles.

B.   Move three toothpicks to make five triangles.

35

Name: _____   Date: _____

Explore space figures

These Space Figures Aren't From Mars

A space figure has three dimensions: length, width, and height.

A.   Is a circle a space figure? _____

B.   Is a rectangular prism a space figure? _____

C.   Is a cone a space figure? _____

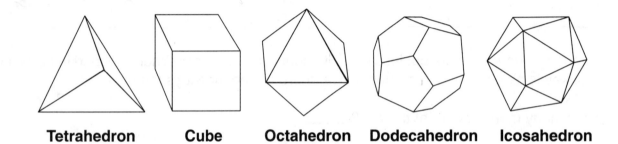

Tetrahedron   **Cube**   **Octahedron**   **Dodecahedron**   **Icosahedron**

Faces are the flat surfaces that form a space figure.

D.   Which three space figures shown have triangular faces? _____

_____

E.   Which space figure shown has square faces? _____

F.   Which space figure shown has faces in the shape of a pentagon? _____

G.   How many faces does a cube have? _____

H.   How many faces does a tetrahedron have? _____

I.   Give two examples of items that are cubes. _____

An edge is the intersection of two faces of a space figure.

J.   How many edges does a cube have? _____

K.   How many edges does a tetrahedron have? _____

L.   How many edges does a dodecahedron have? _____

The points where edges meet on a space figure are called vertices. (singular = vertex)

M.   How many vertices does a cube have? _____

N.   How many vertices does a tetrahedron have? _____

Name: _____   Date: _____

Build a model

Explore an Octahedron

Directions: Create an octahedron. Copy the drawing on heavy paper. Fold all lines backwards and attach with tape.

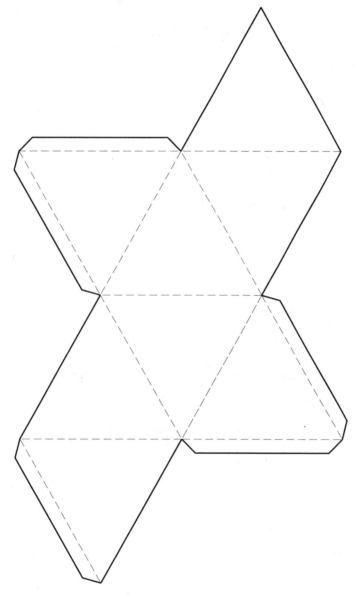

Study your model.

A.   How many faces does an octahedron have? _____

B.   How many edges does an octahedron have? _____

C.   How many vertices does an octahedron have? _____

Name: _____ Date: _____

Compare space figures/Build a model

Triangular Space Figures

A triangular prism has two faces that are congruent polygons. Congruent figures are exactly the same size and shape.

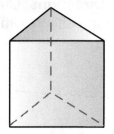

Triangular Prism

A.   What shape are the two congruent faces? _____

B.   How many faces in all? _____

C.   How many edges? _____

D.   How many vertices? _____

Compare the triangular pyramid and the rectangular pyramid.

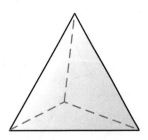

Triangular Pyramid

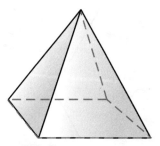

Rectangular Pyramid

	Faces	Edges	Vertices
E. triangular pyramid	_____	_____	_____
F. rectangular pyramid	_____	_____	_____

Imagine what one of these three space figures (triangular prism, triangular pyramid, rectangular pyramid) would look like if it were opened up and laid out flat.

G.   Make a rough draft on scrap paper. When you are certain how it would look, draw it on heavier paper, cut it out, and tape it together to make a model of the space figure.

Name: _____ Date: _____

Compare prisms/Find surface area

Meet Some Prisms

Directions: Fill in the chart.

Prism	Faces	Edges	Vertices	Shape of Base
A. Rectangular prism	_____	_____	_____	_____
B. Pentagonal prism	_____	_____	_____	_____
C. Hexagonal prism	_____	_____	_____	_____
D. Octagonal prism	_____	_____	_____	_____

E. Explain how you could find the surface area of a rectangular prism. _____

_____

F. Would the same method work for finding the area of other prisms? If not, why not?

_____

_____

Name: _____　Date: _____

Match terms with wacky definitions

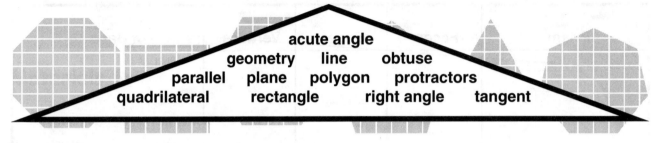

Geometry Riddles

Directions: Use the words in the triangle to find the answers for this wacky geometry quiz.

acute angle
geometry　line　obtuse
parallel　plane　polygon　protractors
quadrilateral　　rectangle　　right angle　　tangent

A.　What's another name for a broken angle? _____

B.　What did the man say when he saw the parrot's cage was empty? _____

C.　What would be a good word for a man who spent the summer at the beach? _____

D.　What did the acorn say when he grew up? _____

E.　What do you call an adorable angle? _____

F.　What geometric figure is a fierce beast? _____

G.　What do you call two L's? _____

H.　What do you call people who are in favor of tractors? _____

I.　What could you call an angle that isn't very smart? _____

J.　What could you call an angle that knows all the answers? _____

K.　What geometry term can you fly? _____

L.　What would you call four lateral passes thrown by a quarterback? _____

Write a riddle using any math term as the answer.

_____

_____

_____

Trade riddles with a partner and solve.

Teacher/Parent page

Math Charades

To play math charades, cut apart the words on this page and the next. Fold each piece of paper in half, and put them in a box or bag. Use the blank slips for words of your choice.

Let students take turns drawing one slip of paper from the box or bag and acting out the word or phrase without using any words or making any drawings.

The one who guesses the answer can take the next turn.

ACUTE ANGLE	ADDEND	ADDITION	ALGEBRA
ARC	AREA	AVERAGE	BISECT
CENT	CENTIMETER	CIRCLE	CIRCUMFERENCE
COMMON FACTOR	CONE	CONGRUENT	COORDINATES
CUBE	DECIMAL	DEGREES	DENOMINATOR
DIAMETER	DIME	DISTANCE	DIVISION
DOLLAR	DOZEN	EDGE	EQUATION
EQUILATERAL TRIANGLE	ESTIMATE	EXPONENT	FACTORS
FORMULA	FRACTION	GALLON	GRAPH
HEIGHT	HEXAGON	HOUR	INCH
INTEGER	KILOGRAM	KILOMETER	LENGTH
LINE SEGMENT	LITER	METER	MILE
MILLIMETER	MINUTE	MULTIPLICATION	NEGATIVE NUMBER

NICKEL	NUMBER LINE	NUMERATOR	OBTUSE ANGLE
OCTAGON	OCTAHEDRON	ONE-FOURTH	ONE-HALF
ORDERED PAIR	OUNCE	PARALLEL LINES	PARALLELOGRAM
PENNY	PENTAGON	PERCENT	PERIMETER
PERPENDICULAR LINES	PI	PINT	PLANE
POLYGON	POSITIVE NUMBER	POUND	PRISM
PROBABILITY	PRODUCT	PYRAMID	QUART
QUARTER	RADIUS	RATIO	RAY
RECTANGLE	RHOMBUS	SECOND	SEVEN-EIGHTHS
SOLUTION	SPHERE	SQUARE	SUBTRACTION
SUM	SYMMETRY	THREE-SIXTHS	THREE-TENTHS
THREE-TWELFTHS	TIME	TRAPEZOID	TRIANGLE
TWO-THIRDS	VARIABLE	VERTEX	VOLUME
WEEK	WEIGHT	YEAR	ZERO

Name: _____ Date: _____

Review Math Terms

Mega-Puzzle: Word Search

Directions: Look up, down, backward, forward, and diagonally to find and circle the 62 math words hidden in the puzzle.

```
R  E  D  N  I  L  Y  C  Q  U  O  T  I  E  N  T
A  A  O  T  N  E  P  O  U  N  C  E  T  I  W  N
T  T  D  X  H  O  M  E  A  B  N  L  N  E  O  E
I  L  A  I  I  R  I  I  R  P  E  E  L  N  N  U
O  O  B  T  U  S  E  T  T  C  V  V  C  O  H  R
T  R  A  Y  A  S  A  E  C  R  E  E  N  T  O  G
H  H  A  L  A  D  U  F  P  A  S  N  G  O  W  N
T  R  G  P  L  R  I  M  O  U  R  N  T  E  T  O
D  E  L  I  M  V  B  Y  L  O  E  F  Q  P  N  C
I  F  N  T  E  A  R  E  Y  L  T  U  E  A  I  P
W  E  O  L  C  T  D  O  G  S  A  N  E  E  P  Y
H  L  D  U  E  O  M  N  O  L  T  M  M  R  L  R
P  D  T  M  R  A  A  E  N  A  A  I  I  H  A  A
A  E  O  D  R  E  E  R  G  E  D  M  C  C  N  M
R  E  E  G  A  L  L  O  N  C  E  N  I  N  E  I
G  R  E  B  M  U  N  A  R  E  A  N  E  Z  O  D
```

ACUTE	ADD	ALGEBRA	ANGLE	ARC
AREA	CONGRUENT	CUBE	CYLINDER	DATA
DECIMAL	DEGREE	DIME	DOZEN	EIGHT
ELEVEN	EQUAL	FIVE	FOOT	FOUR
FRACTION	GALLON	GEOMETRY	GRAM	GRAPH
LENGTH	LINE	MEAN	MILE	MULTIPLY
NINE	NUMBER	OBTUSE	ONCE	ONE
ORDER	OUNCE	PENTAGON	PERCENT	PINT
PLANE	PLOT	POLYGON	PRIME	PYRAMID
QUART	QUOTIENT	RADIUS	RATIO	RAY
ROOT	SEVEN	SIX	SUM	TEN
THREE	TIME	TON	TWELVE	TWO
WIDTH	YARD			

43

Answer Keys

Historical Math (p. 4)
A. $300   B. $1,560   C. 69.5%
D. $832   E. 15,583.5 miles
F. Approximately $10.45 per person

Tackle This (p. 5)
A. Offense: 2,584 pounds; Defense: 3,208 pounds
B. Offense: 235 pounds; Defense: 292 pounds
C. Offense: 6′ 2″; Defense: 6′ 4″
D. Defense   E. 57 pounds
F. Defense   G. 2 inches

Pascal's Triangle (p. 6)

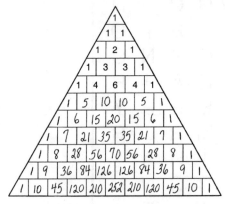

Brrr! (p. 7)
A. 1. Minneapolis (-21), 2. Detroit (-14),
   3. Boston (-11), 4. Chicago (-6),
   5. Pittsburgh (3)
B. <   C. >   D. >   E. <   F. >   G. >
H. -7 -6 0 2 3 4
I. -12 -9 -8 0 1 4 6
J. -8 -6 -3 2 4 5 7
K. -3 -5 -7
L. 1 3 5
M. -10 -15 -20
N. 0 5 10

Help Wanted (p. 8)
Abby $462.25   Ben $650.00
Carlos $719.63   Dana $484.93
Eduardo $630.19
A. 44   B. $512.88 C. $2,027.04

D. and E. Answers will vary. Option 2 allows workers to earn more money per two-weeks. Option 1 requires longer work days, but more days off.

Body Math (p. 10)
A. 72   B. Answers will vary.
C. Answers will vary.
D. 168 miles   E. 7 miles   F. 32 months
G. 292 months (24.3 years)
H. 1,000 weeks (19.2 years)

Surveys Show (p. 12)
A. 12   B. 15   C. 23
D.

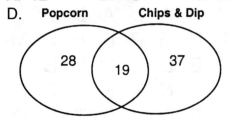

E. 19

F.

Owned Black Shoes / Owned Running Shoes
33   20   47

Music to My Ears (p. 16)
A. 5%, 0.05, $\frac{1}{20}$ B. 6%, 0.06, $\frac{3}{50}$
C. 1%, 0.01, $\frac{1}{100}$ D. 13%, 0.13, $\frac{13}{100}$
E. 11%, 0.11, $\frac{11}{100}$ F. 7%, 0.07, $\frac{7}{100}$
G. 58%, 0.58, $\frac{29}{50}$ H. 30%, 0.3, $\frac{3}{10}$
I. 12%, 0.12, $\frac{3}{25}$

Party Time (p. 17)
Rob	Carlson	Cleaned up	Green
Tina	O'Brien	Invitations	Blue
Maria	Edwards	Decorated	Purple
Carlos	Juarez	Made food	Red

44

Decisions, Decisions (p. 19)
A. 36
B. Each column has one less option than the previous column.

Stop and Smell the Roses (p. 20)
There are 24 possible choices.

Brain Teasers (p. 21)
A. 49 + 50 + 51
B. Scott is 6′ 9″ and Tyrone is 6′ 7″.
C. Harry is 6, Henry is 2, and Hannah is 8.
D. 3 school buses
E. Bus 1: 32, Bus 2: 36, Bus 3: 42.
   Total: 110 passengers
F. Jason is 16. (His grandfather is 64.)
G. Combinations: 12

What's My Line? (p. 22)
A. Possible answers: $\overline{BA}$, $\overline{BC}$, $\overline{DB}$, $\overline{DE}$, $\overline{EF}$, $\overline{HD}$, $\overline{GH}$, $\overline{HI}$, $\overline{IJ}$, $\overline{JK}$, $\overline{KL}$, $\overline{KF}$, $\overline{FC}$
B. Possible answers: $\overleftrightarrow{BC}$, $\overleftrightarrow{DF}$, $\overleftrightarrow{GL}$, $\overleftrightarrow{HA}$, $\overleftrightarrow{CK}$
C. Possible answers: $\overrightarrow{BA}$, $\overrightarrow{DA}$, $\overrightarrow{HG}$, $\overrightarrow{FC}$, $\overrightarrow{FK}$, $\overrightarrow{EF}$, $\overrightarrow{ED}$, $\overrightarrow{IG}$, $\overrightarrow{IL}$, $\overrightarrow{JL}$, $\overrightarrow{KL}$
D. Intersecting   E. Perpendicular
F. Parallel   G. Intersecting

Around and Around (p. 23)
A. 23.6 in.   B. 6,780.2 mi. C. 1,079.7 mi.
D. 785 ft.   E. 2,160   F. 1.5 mi.
G. 1,029.9 ft. H. 244.9 ft. I. 69.1 in.
J. 9.5 ft.   K. 317.1 ft.

Let *t* Represent Temperature (p. 24)
A. $t + 21$ B. $6c$ or $c * 6$
C. $n + 2$ D. $w + 1$, $w + 2$, $w + 3$
E. $s - 20$ F. $h + 0.25h$ or $h + (0.25 * h)$
G. $s * 7$ or $7s$ H. $m * 0.15$ or $0.15m$

To Tell the Truth (p. 25)
A. $7 + x = x + 7$ or $5 + x = x + 5$
B. $98 + (-48) = 50$ or $98 + (-46) = 52$
C. $15 * 4 = 60$ or $20 * 3 = 60$
D. $\frac{7}{3} = 2\frac{1}{3}$ or $\frac{5}{2} = 2\frac{1}{2}$
G. $9 + 5 = 14$

H. $(14 + 21) + 7 * 6 = 77$
I. $\dfrac{(54 - 6) * 2}{8} = 12$
J. $(7.3 * 3) - (6.4 * 2.1) = 8.46$
L. A true equation is a mathematical statement in which both sides are equal.

Triangle Tango (p. 27)
A. and B. This is an example of a right triangle. Drawings may vary.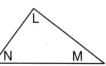
C. Answers may vary. At least one angle must be 90°.
D. 180°
E. and F. This is an example of an acute triangle. Drawings may vary. All angles must be less than 90°.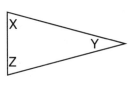
G. Answers will vary.
H. 180°
I. and J. This is an example of an obtuse angle. Drawings may vary. One angle must be greater than 90° and less than 180°.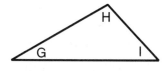
K. Answers will vary.
L. 180°

World Records (p. 28)
A. 122.66 ÷ 2 = 61.33 ft. (radius)
   61.33^2 x 3.14 = 11,810.7 sq. ft.
B. 11,810.7 - 10,057 = 1,753.7 sq. ft.
C. 134° - (-79.8°) = 213.8°
D. 40 x 23 = 920 sq. ft.
E. 40 + 40 + 23 + 23 = 126 ft.
F. 172 ft. x 12 in. + 4 in. = 2,068 in.
G. 2,068 in. ÷ 36 in. = 57.4 yd.

More Triangles (p. 31)
A. No. The three angles of an equilateral triangle are equal. A right triangle has one angle of 90°. Since the sum of the three angles equals 180°, a triangle with three 90° angles is impossible.

B. Yes. Drawings may vary. Here is one example.

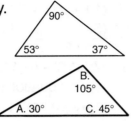

C. Drawings may vary. Here is one example.

D. Here is an example.

Swim Ballet (p. 32)

A. The minimum needed would be 36 swimmers: a solid triangle with 8 on each side and a solid square with 6 rows of 6.
B. Yes. Because 81 is the product of 9 x 9.
C. One solution

```
XX   XX   XXX    XXXXXXXX
XX   XX   XXX    XXXXXXXX
          XXX    XXXXXXXX
                 XXXXXXXX
                 XXXXXXXX
                 XXXXXXXX
                 XXXXXXXX
                 XXXXXXXX
```

D. No: 3, 6, 10, 15, 21, 28, 36, 45, 55, 66, 78, 91, etc., are triangular numbers. 81 is not a triangular number.

Think It Through (p. 35)

A. 21, 34, 55 (Each number after the second is the sum of the two numbers before it. This is called a Fibonacci sequence.)
B. The only possible combination of coins Jake could have are four pennies, two dimes, and two quarters. He could not lend her exactly 30 cents.
C. 40
D. A blue boomadinger costs $40 and a red one costs $50.
E. $80
F. He had 45 $5 bills and 50 $1 bills.

Shape Up (p. 35)

A.

B.

These Space Figures Aren't From Mars (p. 36)

A. No
B. Yes
C. Yes
D. Tetrahedron, octahedron, and icosahedron
E. Cube
F. Dodecahedron
G. 6
H. 4
I. Answers will vary
J. 12
K. 6
L. 30
M. 8
N. 4

Explore An Octahedron (p. 37)

A. 8
B. 12
C. 6

Triangular Space Figures (p. 38)

A. triangular
B. 5
C. 9
D. 6
E. 4; 6; 4
F. 5; 8; 5

Meet Some Prisms (p. 39)

A. 6; 12; 8; rectangle
B. 7; 15; 10; pentagon
C. 8; 18; 12; hexagon
D. 10; 24; 16; octagon
E. The surface area of a rectangular prism is the sum of the area of all its faces.
F. Yes

Geometry Riddles (p. 40)

A. rectangle
B. polygon
C. tangent
D. geometry
E. acute angle
F. line
G. parallel
H. protractors
I. obtuse
J. right angle
K. plane
L. quadrilateral

Mega-Puzzle (p. 43)

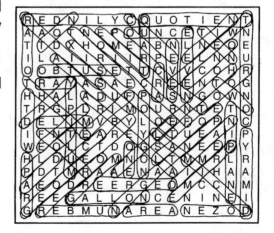